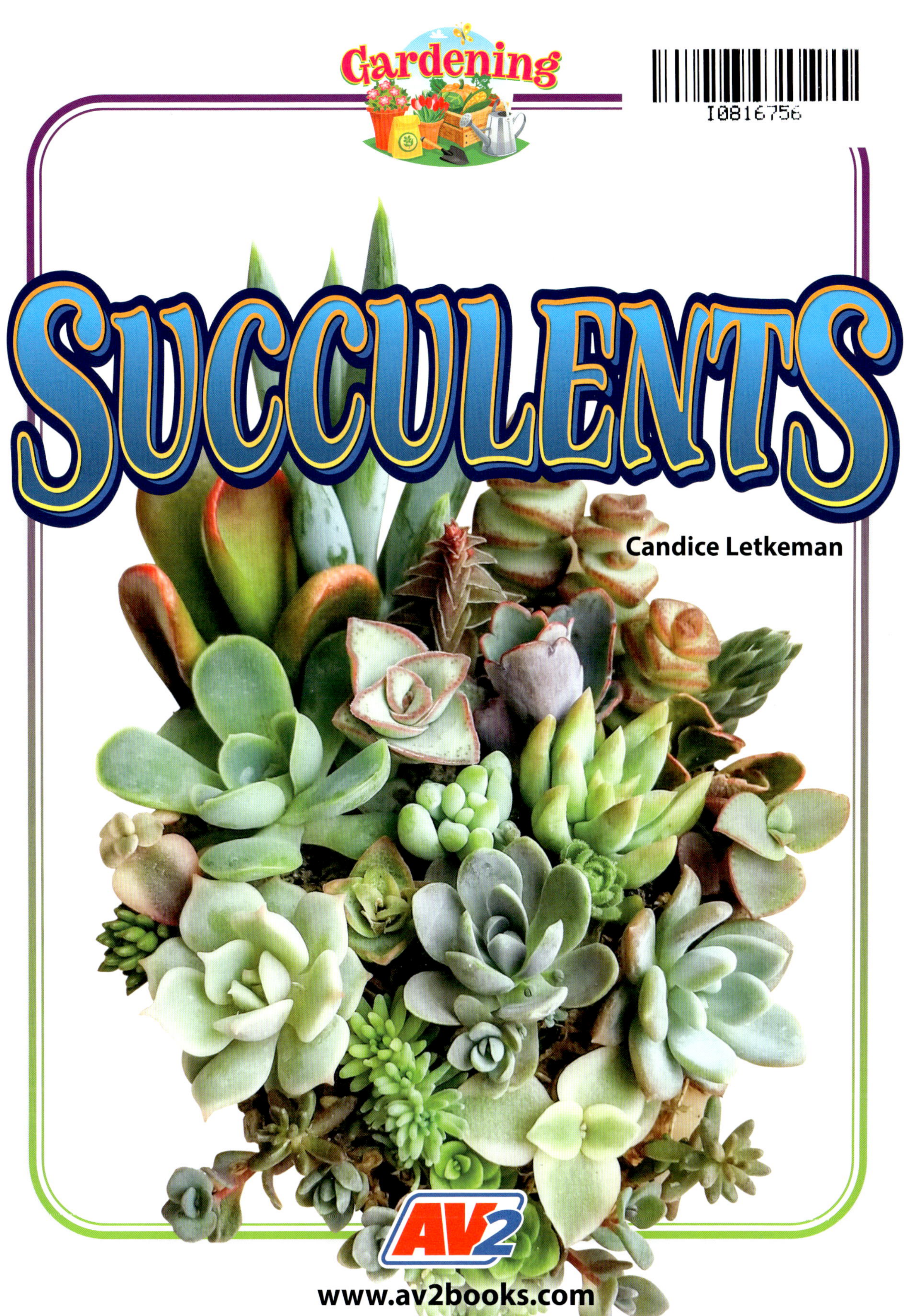

SUCCULENTS

Candice Letkeman

AV2

www.av2books.com

Step 1
Go to **www.av2books.com**

Step 2
Enter this unique code

RBHGO87PG

Step 3
Explore your interactive eBook!

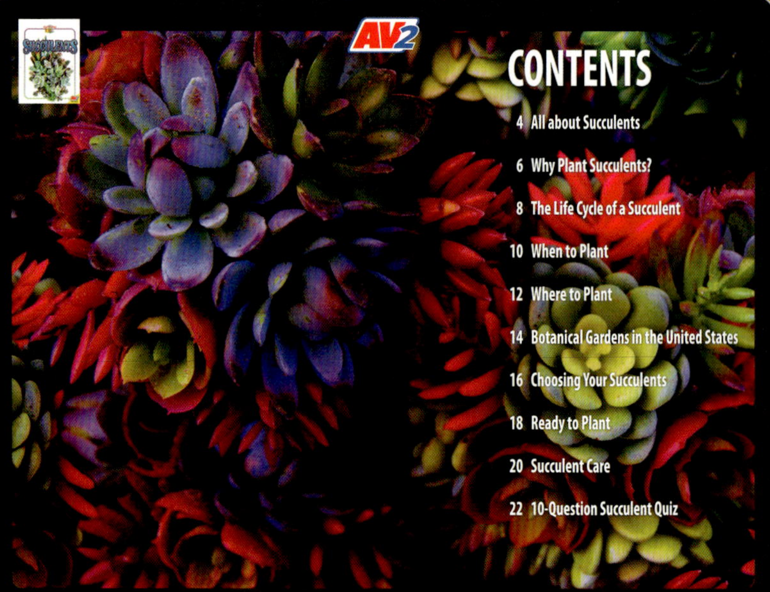

AV2 is optimized for use on any device

Your interactive eBook comes with...

Contents
Browse a live contents page to easily navigate through resources

Audio
Listen to sections of the book read aloud

Videos
Watch informative video clips

Weblinks
Gain additional information for research

Try This!
Complete activities and hands-on experiments

Key Words
Study vocabulary, and complete a matching word activity

Quizzes
Test your knowledge

Slideshows
View images and captions

This title is part of our AV2 digital subscription

1-Year K–5 Subscription
ISBN 978-1-7911-3320-7

Access hundreds of AV2 titles with our digital subscription.
Sign up for a FREE trial at **www.av2books.com/trial**

SUCCULENTS

CONTENTS

All about Succulents

If you were to grow a succulent garden with your mom or dad, what types of plants would you put in it? Why did you choose these ones? Do you like the way they look? Will you keep your succulents indoors or outdoors?

Growing succulents is a fun way to learn about plants. You can take care of them every day. You will find out what your succulents need to grow and be healthy. Having succulents in your home can also help keep you healthy!

Cacti are one of the largest succulent families. There are more than 2,000 different types of cactus.

What Is a Succulent?

A succulent is a plant that stores water in its thick, fleshy **tissues**. Succulents have several parts.

Spines

Some succulents have sharp spines. They help protect the plant from animals.

Leaves

Many succulents use their leaves for **photosynthesis**. The leaves are also used to store water.

Roots

Roots are spread out underground. They help the succulent absorb water.

Stem

For plants without leaves, such as cacti, photosynthesis occurs in the stem. These plants also store water here.

Succulents are found on every continent except Antarctica.

The old man cactus is named for its white hair. It can live for up to 200 years.

Why Plant Succulents?

Succulents add beauty to outdoor gardens. They can also liven up indoor spaces. These plants are important for other reasons as well.

Succulents help clean the air. They take in harmful indoor **pollutants** and change them to make food for the plant. Succulents also release **water vapor**, which adds moisture to the air. This can be good for someone with a sore throat or dry skin.

Agave is a natural sweetener. Some people use it in place of sugar.

At night, succulents release **oxygen** into the air. This can help people sleep better.

Gel from the leaves of **aloe vera** plants can treat sunburns.

The Life Cycle of a Succulent

All succulents begin life, grow, and make more succulents. This is the life cycle of a succulent.

Seeds

Succulents begin life as seeds. A seed will grow once it is planted.

Fruit

Some succulent fruits are **edible**. These fruits have seeds in them. Birds and other animals eat the fruit and spread the seeds. The cycle begins again.

Roots

Roots develop and hold the plant in the ground.

Stem

The stem grows above ground. Some succulents grow leaves. Cacti develop arms.

Flowers

Some succulents grow flowers. Insects and other animals **pollinate** the flowers. Then, the flowers become fruit.

When to Plant

What is the weather like where you live? Succulents need certain conditions in order to grow well. They must be planted at the right time of year. If it is too cold when you plant them, your succulents may not grow.

Most succulents only grow outside in places where the ground never freezes. If you are planting your succulents outside, make sure to research whether or not the plants you choose will be able to grow in your region. You can grow succulents indoors wherever you live.

SPRING

Spring is the best time to plant succulents. This is when they are actively growing.

SUMMER

Succulents establish a healthy root system quickly in warm summer temperatures.

WINTER

Some types of succulents can grow in winter temperatures of –20° Fahrenheit (–29° Celsius).

Baobab trees are the **world's largest** succulents. They can grow up to **82 feet high**. (25 meters)

Some cacti can go **without water** for up to **two years**.

Where to Plant

It is important that you choose the right place to plant your succulents. How do you know if a place is right? There are three main factors you should consider.

Sun

Your plants should be somewhere that gets sunlight. Most succulents grow best in direct sun, although some also need shade. Without enough light, succulents may become **leggy** and weak.

Water

All plants need water to survive. However, succulents need less water than other kinds of plants. Most succulents need water only when the soil is completely dry. Choose a spot with little rainfall.

Soil

Succulents need soil with good **drainage**. This helps keep their roots dry. If the soil holds water for too long, your succulents will not grow.

Botanical Gardens in the United States

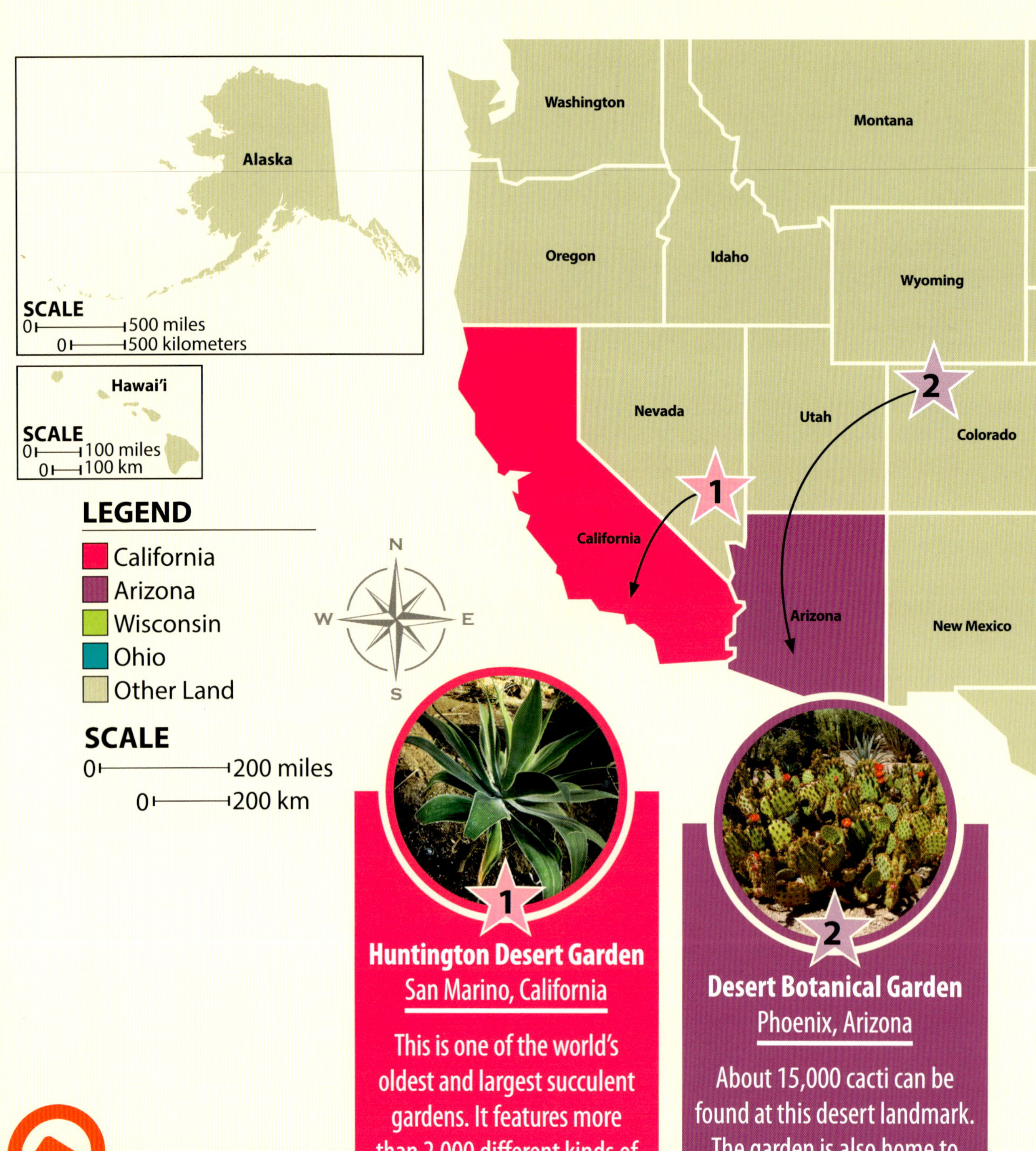

Huntington Desert Garden

San Marino, California

This is one of the world's oldest and largest succulent gardens. It features more than 2,000 different kinds of succulents and desert plants.

Desert Botanical Garden

Phoenix, Arizona

About 15,000 cacti can be found at this desert landmark. The garden is also home to 186 different kinds of agave.

There are **botanical gardens** across the United States. These gardens have several different types of plants. Some botanical gardens feature succulents from around the world. People can visit these gardens to enjoy the plants and learn more about them.

Mitchell Park Domes

Milwaukee, Wisconsin

Visitors to the Domes can see succulents such as bunny ears cacti. These unusual plants get their name from their unique shape.

Franklin Park Conservatory and Botanical Gardens

Columbus, Ohio

This conservatory has a multi-leveled Desert Biome. It features a variety of succulents and cacti blooms.

Choosing Your Succulents

There are so many types of succulents in the world. How do you decide which ones to plant? One of the first things you have to think about is where you want to grow your succulents. Are they going to be part of an outdoor or an indoor garden? If your succulents are being grown outside, a **xeriscape** might be a good choice for you. If you are planting inside, consider a tray garden instead.

Succulents are interesting to look at. They come in many different shapes, sizes, and colors.

Xeriscapes

Xeriscapes can showcase the unique shapes and colors of succulents. Large succulents make good xeriscaping plants because they are outside, where they have plenty of room to grow.

Tray Gardens

The shallow containers used in tray gardens allow you to create tiny indoor landscapes. Small succulents grow well in these gardens because they often have shallow root systems.

Ready to Plant

If you want to grow a successful garden, you have to start with the right equipment. You also have to follow the planting process.

Gardening Equipment

Planting Trays

If you are growing different kinds of succulents, use planting trays to keep each type separate. The trays should be no more than 4 inches (10 centimeters) deep.

Plastic Wrap

Use plastic wrap to cover your planting trays. This keeps the soil moist until the seeds **germinate**.

Toothpick

Use a toothpick to gently spread your succulent seeds. You can also poke holes in the plastic wrap to provide some air **circulation**. This will help keep your plants free from pests and diseases.

The Planting Process

1

Fill your planting tray with soil. You can use soil that has been prepared just for succulents, or mix regular soil with sand.

2

Dampen the surface of the soil with water so the seeds will stick to it.

3

Carefully spread the seeds on top of the soil, using your toothpick to separate them. Do not cover the seeds with soil.

4

Cover the tray with your plastic wrap and poke a few holes in it with the toothpick.

5

Place the tray somewhere bright, out of direct sunlight. Keep the soil moist, but not wet.

Repeat for each kind of succulent you want to plant.

Succulent Care

Taking care of a garden does not stop with planting. You must continue to check on your succulents to make sure that they are growing and staying healthy. There are three main areas that could need attention.

Fertilizer

Adding **fertilizer** can help your succulent garden grow. How often you fertilize depends on the size, type, and growing conditions of your succulents.

Water

Adding too much water or watering too often can kill your succulents. Make sure that the soil is completely dry for a few days between waterings.

Protection

If it rains, you may need to cover your plants. Most succulents cannot survive **frost** either. If your plants are in pots, bring them indoors during heavy rain or freezing temperatures.

10 Question Succulent Quiz

1 Why do succulents need soil with good drainage?

2 How do succulents begin life?

3 What kind of succulent gardens are a good choice for planting outdoors?

4 For how long can the old man cactus live?

5 In which botanical garden can you see bunny ears cacti?

6 What do succulents release into the air that can help people sleep better?

7 Why should you cover your planting trays with plastic wrap?

8 Where do cacti store water?

9 What can gel from the leaves of aloe vera plants treat?

10 When do most succulents need water?

ANSWERS 1. To keep their roots dry 2. As seeds 3. Xeriscapes 4. Up to 200 years 5. Mitchell Park Domes in Milwaukee, Wisconsin 6. Oxygen 7. To keep the soil moist until the seeds germinate 8. In their stems 9. Sunburns 10. When the soil is completely dry

Key Words

aloe vera: a succulent that produces a gel used in health and cosmetic products

botanical gardens: places where plant research takes place

circulation: the act of moving or flowing through a particular course

drainage: the removal of water from soil

edible: able to be eaten

fertilizer: a mixture that adds nutrients to soil

frost: a thin coating of ice that forms in spring or fall

germinate: begin to grow and put out shoots

leggy: having a long, thin stem and small leaves

oxygen: a chemical element found in the air as a colorless, odorless, and tasteless gas that is necessary for life

photosynthesis: the process by which plants use sunlight to make food

pollinate: to carry plant dust from one plant to another so seeds can be made

pollutants: substances that make the air or water impure or unsafe

tissues: a group of cells in a plant or animal that are similar in form and function

water vapor: water in the form of a gas

xeriscape: a garden that does not need much water and is well suited for hot areas

Index

Published by AV2
276 5th Avenue
Suite 704 #917
New York, NY 10001
Website: www.av2books.com

Library of Congress Cataloging-in-Publication Data
Names: Letkeman, Candice, author.
Title: Succulents / Candice Letkeman.
Description: New York, NY : AV2, [2022] | Series: Gardening | Includes index. | Audience: Grades 2-3
Identifiers: LCCN 2020022373 (print) | LCCN 2020022374 (ebook) | ISBN 9781791127794 (library binding) | ISBN 9781791127800 (paperback) | ISBN 9781791127817 (ebook other) | ISBN 9781791127824 (ebook other)
Subjects: LCSH: Succulent plants--Juvenile literature.
Classification: LCC SB438 .L47 2021 (print) | LCC SB438 (ebook) | DDC 635.9/525--dc23
LC record available at https://lccn.loc.gov/2020022373
LC ebook record available at https://lccn.loc.gov/2020022374

Printed in Guangzhou, China
1 2 3 4 5 6 7 8 9 0 25 24 23 22 21

022021
101120

Editor: Katie Gillespie
Art Director: Terry Paulhus

Every reasonable effort has been made to trace ownership and to obtain permission to reprint copyright material. The publisher would be pleased to have any errors or omissions brought to its attention so that they may be corrected in subsequent printings. AV2 acknowledges Getty Images, Alamy, Dreamstime, iStock, and Shutterstock as its primary image suppliers for this title.